Space Elevator Survivability
Space Debris Mitigation

International Space Elevator Consortium
Fall 2010

Peter Swan
Robert "Skip" Penny
Cathy Swan

A Study for Progress in
Space Elevator Development

Space Elevator Survivability
Space Debris Mitigation

Copyright © 2015 by:
International Space Elevator Consortium

Published by Lulu.com
dr-swan@cox.net

ISBN 978-1-329-09392-8

Printed in the United States of America

Preface

A yearly topic is selected to concentrate the International Space Elevator Consortium efforts toward a meaningful goal or objective. This focus enables the community to contribute towards the designated goal and participate in various activities such as ISEC Journal articles, student contest topics, conference themes, and major studies.

This study represents the culmination of efforts by the contributors and answers the question:

Will space debris be a "show stopper" for the development of the Space Elevator Infrastructure?

The answer is a resounding NO!

The recognition of space debris risk with reasonable probabilities of impact is an engineering problem. The proposed mitigation concepts change the issue from a perceived problem to a concern; but, by no means is it a significant threat. This study illustrates how the development office for a future space elevator infrastructure can attack this problem and convert it into another manageable engineering problem.

Ted Semon

President, ISEC

Space Elevator Survivability
Space Debris Mitigation

Executive Summary

The Big Sky theory of Space Debris, or the "what, me worry?" approach, has faded into the past as have Sputnik and the Saturn rocket. The space community now recognizes that the continuous growth (chart below[1]) of objects remaining in orbit has lead to an arena where space debris mitigation and removal becomes mandatory. Indeed, the Space Elevator community is concerned about space debris numbers and densities because of its dramatic growth over the last three years. During the study, the team addressed many issues to include:

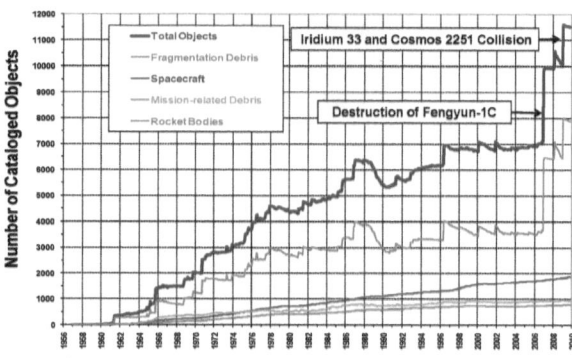

- The probabilities of collision in Low Earth Orbit (LEO), in Geosynchronous Earth Orbit (GEO), and in Medium Earth Orbit (MEO).
- The growth rate as it threatens an operational space elevator.
- A reasonable approach for space elevator developers to ensure infrastructure safety.
- Approaches to interrupt sources of debris.

---

[1] *With permission from Debra Shoots, NASA Orbital Debris Program Office, May 2010.*

- Mitigation of risk for the space elevator community through design, operations, policies, and lowering the threat.

To assess the risk to a space elevator, we have used methodology from the 2001 International Academy of Astronautics (IAA) Position Paper on Orbital Debris[2]:

"The probability (PC) that two items will collide in orbit is a function of the spatial density (SPD) of orbiting objects in a region, the average relative velocity (VR) between the objects in that region, the collision cross section (XC) of the scenario being considered, and the time (T) the object at risk is in the given region."

$$PC = 1 - e^{(-VR \times SPD \times XC \times T)}$$

Using this formula, we calculate the Probability of Collision for LEO, MEO, and GEO. Our focus is on LEO -- as fully two thirds of the threatening objects are in the 200-2000 km (LEO) regime. Our analyses show:

The threat from Space Debris can be reduced to manageable levels with relatively modest design and operational "fixes."

Our hope is that this study will raise the awareness of the problem (and spur action to implement policies and directives to mitigate and reduce the risk of collision) to the space elevator stakeholders and all other users of the near Earth space environment.

---

[2] 2001 Position Paper On Orbital Debris, International Academy of Astronautics, 24.11.2000.

Chapter 1 – Introduction

1.0   General Background

Orbital debris will pose a hazard to a space elevator. The 100,000 km long, one meter wide ribbon includes this as a potential vulnerability. To establish a space elevator program, the issue of space debris must be addressed through the establishment of requirements for debris tracking and estimating locations, both current and projected, and "rules of the road" for debris mitigation and removal. Derivative requirements such as space elevator segment location, response time, and anchor platform maneuverability must also be addressed. This pamphlet will assess the risk of debris damage to a space elevator, present potential mitigation measures, and make recommendations with respect to a space elevator and the space debris environment.

The modern day space elevator, as described by Dr. Edwards in *Space Elevators*[3], is the future for space access; however, understanding the environment in which it will operate is paramount to its success. As outlined in *Space Elevator Systems Architecture*[4], there are many threats to a space elevator; however, for each threat there are multiple engineering mitigation techniques. This pamphlet will address one such threat, describe the magnitude of concern, and then suggest mitigation techniques. When considering space debris and its threat to a space elevator, some significant questions have to be asked:

- Does space debris cause concern for space elevator developers?
- How precisely does one need to know the location of space elevator ribbon segments?

---

[3] Edwards, Bradley C. and Eric A. Westling, The Space Elevator. BC Edwards, Houston, TX, 2003.
[4] Swan, Peter A. and Cathy W. Swan, Space Elevator Systems Architecture, Lulu.com, 2007.

- How precisely does one need to know the location, and propagated location of large space debris?
- What are the projected levels of concern and what needs to be accomplished prior to operations?
- How do we mitigate the risk of orbiting debris and satellite collisions with the space elevator?
- What is the probability of puncture from impacts with small debris?
- What is the probability of a sever by large orbiting objects?

This study will discuss multiple altitude regions, ranging from LEO, where the greatest hazards exist, to beyond GEO, where micrometeoroids are the primary threat. Research addressed three debris threat categories: (a) small (less than 10 cm), which are numerous with random direction: (b) tracked and inert (10 cm and larger), with known numbers and orbital characteristics; and, (c) large and controllable active satellites (about 6% of tracked objects). The combination of region (as defined by altitude) and threat types (as defined by size) will enable the reader to understand the environment in which a future space elevator must operate. The quantitative space debris information was gathered from the NASA Office of Space Debris at the Johnson Spaceflight Center.

1.1   Chapter Breakout

This pamphlet is based upon the modern day design of a space elevator as described by Dr. Brad Edwards. The authors realize that there have been many proposals for alternative designs with respect to his approach. As the community does not have a funded program with a need to finalize designs, even more ideas will continue to appear. The authors standardized the design to enable this analysis and presentation of data. As the community narrows in on a "real" system, discussions on space debris will focus on the applicable critical items and final design. This ISEC Pamphlet presents a survivability approach for a space elevator with respect to the threat of space debris. This includes the following chapters:

Chapter 1   Introduction This chapter lays the groundwork for the whole pamphlet. It is a quick discussion of the space elevator concept and a definition of the problem. A presentation of altitude segmentation of the regions in space enables the analyses to proceed by considering the unique aspects of each region.

Chapter 2   Definition of the Problem Space debris status, now and into the future. This is a straightforward presentation of the numbers of man-made objects in the defined altitude regions. A quick discussion of sizes and orbits enables analyses to proceed and an understanding of the problem to be presented.

Chapter 3   Probability of Impact   This chapter presents an approach to calculating these important numbers. Each region has a different set of issues and presents a slightly different set of probabilities. The collision probability for each region is then calculated – which will lead to a better understanding of the criticality of space debris to a successful space elevator project.

Chapter 4   Mitigation Techniques   Any space elevator must be designed and operated within a "safe" environment. The concept of a space elevator infrastructure includes multiple pairs of tethers around the world. Complete severing of all space elevator ribbons would be catastrophic. Therefore, mitigation techniques must ensure that there are multiple strands per ribbon and multiple space elevators always in operation.

Chapter 5   Conclusions This chapter will summarize the various regional threats to a space elevator, and potential mitigation techniques for each threat.

Chapter 6   Recommendations This study will present recommendations that should lead to actions within the development of a space elevator transportation infrastructure. Additionally, this study will present recommendations for the space debris community.

1.2   What is the Space Elevator Concept?[5]

The modern day space elevator has many strengths and is the future for space access. For the purpose of this pamphlet the general characteristics include:

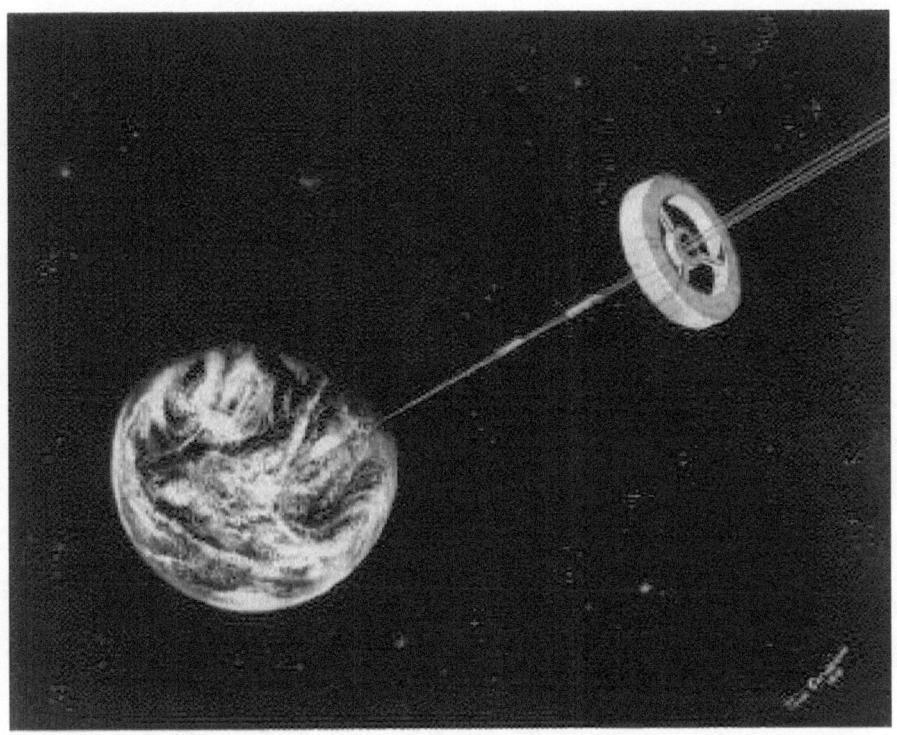

Figure 1.1 Space Elevator from Pearson's work[6]

---

[5] A majority of this section comes from Chapter 2 of Space Elevator Systems Architecture, Swan, Peter & Cathy Swan, Lulu.com publishers, 2007.
[6] Pearson, Jerome. "Space Elevator, a US Air Force Painting." 1975.

- Length: 100,000 km, anchored on the Earth with a large mass floating in the ocean and a counterweight at the apex.
- Width: One meter.
- Ribbon Design: Woven with multiple strands to enable localized damage compensation and curved to ensure edge-on, small sized, hits do not sever the ribbon.
- Cargo: The first few years will have five concurrent ribbon riders [20 tons], each with 20 ton payloads without humans (radiation tolerance is an issue for any two week trip through the radiation belts).
- Operation: It is assumed that multiple space elevators will be in operation by 2030.
- Construction Strategy: The space elevator will initially be deployed from GEO. Once the gravity well has been overcome, it will be replicated from the ground up leading to multiple elevators appearing around the globe. This redundancy will reduce the magnitude of catastrophe if one is lost. We will never be trapped inside Earth's gravity well again!

1.3   Study Approach

Space elevator survivability against space debris can be simplified by approaching each issue as if it were the most critical item and not influenced by the complexity of the project. Simplicity in design is definitely a desirable outcome of early brainstorming for the development of a mega-project. Combining simple concepts leads to more complexity; however, small pieces tend to go together instead of forcing a larger solution up from the bottom. Answers will surface and should be globally applicable. The selected approach for space debris mitigation analyses is along altitude regions. The varying characteristics of different altitude regions drive design requirements in different directions. This segregation seems to be natural and reflects the varying requirements of a space elevator design. The survival aspects of the design will be presented along altitude segregated regions:

beyond geosynchronous, geosynchronous, medium Earth, low Earth, and the aeronautical region.

Table 1.1 Altitude Regions

Region	From (km)	To (km)
Super – GEO	35,880	100,000
GEO	35,680	35,880
MEO	2,000	35,680
LEO	Spaceflight limit (200 km)	2,000
Aero Drag	Sea Level	Spaceflight limit (200 km)

[GEO – geosynchronous orbit @ 35,786 km; MEO-Medium Earth Orbit; LEO – low Earth orbit: radius Earth = 6378 km]

1.4   Altitude Breakout

The rationale for segmenting the space elevator system into altitude regions is based upon simplicity and engineering scope. Solving local problems is always easier than solving global problems. This breakout allows the space systems architect and lead space systems engineer to compare and contrast engineering alternatives across the total project, allowing optimization at the appropriate levels. Obviously, simple approaches inside a region could be expandable to other regions. Hopefully, the insight gained by these analyses will yield an opportunity to lead design concepts and systems alternatives. But first, the following tables compare the altitude regions by basic characteristics and their effects upon design.

Table 1.2 Super GEO (Altitude > 35,880 km)

Design Driver	Impact on Design
Centrifugal force dominates	No power required to leave GEO towards end mass
Low probability of collisions	Backup Simplicity
Launch location for solar system	Flexibility
Grow as counter-weight	Survivability and flexibility
Capture old GEO satellites	"Free mass" for counter-weights

Table 1.3 GEO (35,680 < Altitude < 35,880 km)

Design Driver	Impact on Design
Minimal survivability threat	Simplicity
Dominant during developmental phase	Center of mass and tension measurements
Critical transportation node	Work station (with or without people)
GEO node attach-detach as climbers pass altitude	Understanding of local dynamics and robotic grappling
Maybe GEO node not attached to space elevator – just floats along side	Creative design option needs to be traded

Table 1.4 Medium Altitude (2,000 < Altitude < 35,680 km)

Design Driver	Impact on Design
Self deploy	Minimum design
LEO/MEO satellite nodes	Launch and inclination issues
Real debris issues (Molina, GEO Transfer Orbit, Navigation orbits)	Survivability and redundancy
Electric propulsion probable	Simplicity
Radiation belts - lower region	Dump radiation
Tension monitoring – GPS location	Equipment and communications

Table 1.5 LEO (spaceflight limits 200 km < Altitude < 2,000 km)

Design Driver	Impact on Design
Robust ribbons	Survivability and multiple tracks
Traffic control (up to 2,500 km)	Simplicity
Survivability of space elevator at greatest risk	Safety, security, move ribbon, curved surface, wide ribbon
Large radiation environment	Proper coating to materials Potential lowering of radiation inside electron and proton belts
Hotel for tourists at 100 km	Early revenue and work space
Laser energy efficient	Simplicity

Table 1.6 Aero Drag (sea level to spaceflight limit 200 km)

Design Driver	Impact on Design
Minimum tension at connection	Simplicity and less stress
Multiple up and down paths	Redundancy and traffic management
Redundancy against terrestrial threats	Survivability
Base anchors distributed over large radius	Redundancy and flexibility
Traffic control in Command and Control Center	Local knowledge and flexibility
Lightning mitigation (laser illumination)	Survivability
Deploy prior to connection	Ease of space elevator deployment
Execute when ribbon deployed	Simplicity
Boat horizontal motion drive climbers vertical	Unique propulsion idea

1.5   General Threat Breakout

A systems approach to space elevator survival must address all threats from the expected environments. As such, a quick discussion on the other threats puts space debris in perspective. The threats logically separate into five altitude regions and encompass all basic issues that must be evaluated. This ranges across many arenas, to include:

- Meteors and micrometeoroids
- Space debris (expired spacecraft and/or fragments)
- Operational spacecraft
- Space environment (x-rays, gamma rays, atomic oxygen, charged particles, equilibrium temperatures)
- Atmospheric environment (winds aloft, hurricanes, tornados, lightening, etc.)
- Human environment (aircraft, ships, terrorists, etc.)

Super GEO: There is very little human-created debris in this region, so the major threat consists of meteors and micro-meteoroids.

GEO Region: Problems in this region include the micro-meteorite issue and human hardware intersection. The advantage is that debris is mostly large and moving slowly when in, or close to, the "GEO Belt." The relative velocities are usually less than 10s of meters per second. However, current guidelines for GEO satellite removal call for raising their orbits at least 200 km, and lunar and solar perturbations can cause inclination changes, raising the relative velocities of potential collisions with the space elevator. This leads to the conclusion that most of these dead satellites will have to be removed.

MEO Region: Few man-made objects reside in this region; and in the context of space debris, MEO resembles GEO. There are a small number of objects right above the lower limit of 2,000 km altitude; less than 200 around the 12 hour circular orbit populated by navigation constellations (GPS with more than 36 satellites - GLONASS with more than 20 satellites - and the future Galileo with more than 24 satellites at 20,200 km); and, in addition, the Geosynchronous Transfer Orbit (12 hour, highly elliptical) retains rocket bodies after payloads are "kicked" into GEO orbit. The velocity differences between a space elevator and orbiting objects for this elliptical debris present a serious threat: however, the numbers are small. In addition, the lower portion of this region contains radiation belts.

LEO Region: Low Earth Orbit has a major problem with space debris, a modest problem with operational satellites, and a smaller problem with micrometeoroids. Most catalogued space debris exists in this region, filling all altitudes and inclinations, which results in equatorial crossings near any space elevator. Of the 15,000 objects tracked daily, approximately 12,000 are located in this region. A quick look at the numbers and volume leads to the figure that illustrates the flux of debris vs. dimension.

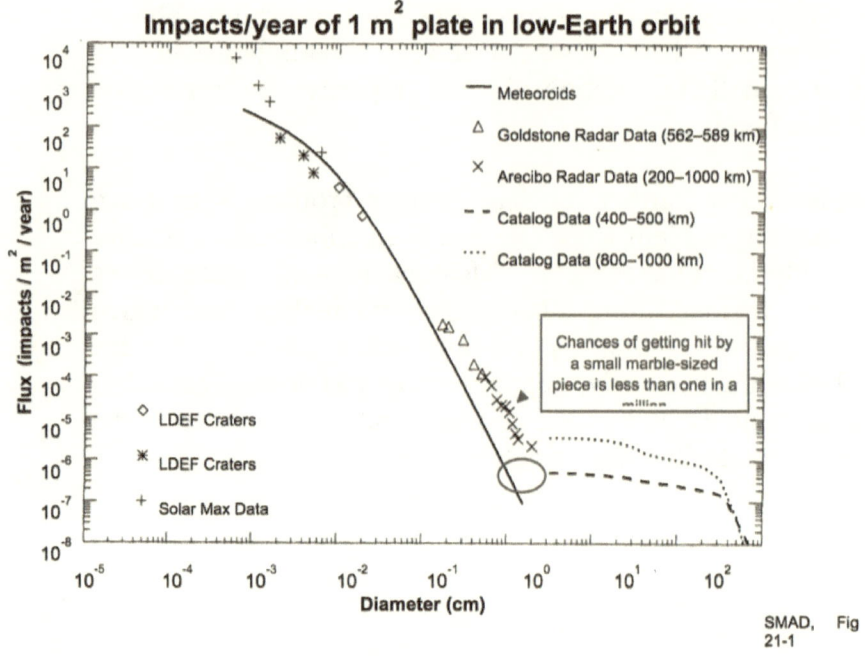

Figure 1.2 Impact Rates for Meteoroids and Orbital Debris[7]

Aero Drag Region: The atmosphere will threaten the ribbon and integrity of the space elevator in this region. The dangers of concern are: winds aloft, hurricanes, tornados, lightening, and human interference (aircraft, ships, and terrorism).

[7] Larson, Wiley and James Wertz, Space Mission Analysis and Design. Ed. III._ McGraw Hill, 2002, p. 841.

Chapter 2 – Definition of the Problem

2.0   Introduction

The primary concern for those studying space debris deals with "what is up there?" Space debris is defined as anything man-made that is in orbit and comes in multiple categories and sizes. There are large rocket bodies as well as large and medium sized spacecraft that are no longer functioning. There are functioning spacecraft -- large, medium and small. And, there are pieces of junk -- large, medium and small. This chapter will lay out a short history of how the global space community slipped into this situation while helping to define the population density and distribution. One key element is the space community's ability to know precisely where space debris objects are at any given time and to predict their locations for the future.

2.1   History of Space Debris

Concern about space debris can be divided into four historical phases:

- Big Sky Theory   (1957-1970)
  No concern because there is so much volume
- What is up There?   (1970-1989)
  Scientists/Military wonder what is up there.
- Collision Concern   (1989-2009)
  Scientists/mathematicians worry about collisions
- Collision Reality   (2009 +)
  The IRIDIUM-Cosmos Fragmentation

2.1.1 Big Sky Theory (1957-1970)

Space debris has long been a thorn in the side of space operations. Exploding rocket bodies and batteries, cameras floating away from astronauts, and old, dead, satellites or rocket bodies all created worthless parts going at orbital velocities. The volume of space surrounding Earth is huge and for many years the issue of space debris was of no concern. During this time period curious astronomers and interested military officers wanted, or needed, to know what was up there and who was doing what. They developed routine systems tracking operational satellites, and any other objects larger than 10 centimeters. Catalogs were established and predictions for rendezvous (collisions) were determined to be very small.

2.1.2 What is up There? (1970-1989)

During this phase, researchers attempted to determine what was really in orbit and to whom it belonged. At this time, geopolitical concerns were paramount, and few people were researching residual junk and where it was going. This research focused on counting and predicting collisions with low probabilities. Initial efforts were focused on lowering future debris by issuing design guidelines. In addition, the permanent presence of humans (with space stations and space shuttles) heightened concerns for safety of flight. At the same time, both the US and USSR conducted anti-satellite (ASAT) tests, resulting in additional space debris.

2.1.3 Collision Concern (1989-2009)

During this phase many scientists and operators projected major concerns for the future; however, very little progress was made to reduce debris in orbit. Much was accomplished in creating guidelines for design of spacecraft and rocket bodies culminating in a document expressing the desire for "zero debris creation" as a goal. Most space faring nations incorporated these rules; but, as the rules were voluntary, there was no mechanism for enforcement. Great strides were being made in calculations predicting future debris populations and the Kessler Cascade theory became generally accepted. [note: Kessler Cascade described in section 2.1.4] During this time, at least eight collisions occurred (see Table 2.1) reinforcing the realization that the situation was changing. In addition, safety of human spaceflight became a serious concern with six permanent residents on the International Space Station.

Table 2.1 Satellite Collisions [Complied by Dr. David Wright Union of Concerned Scientists][8]

Year	Satellites
1991	Inactive Cosmos 1934 satellite hit by cataloged debris from Cosmos 296 satellite
1996	Active French Cerise satellite hit by cataloged debris from Ariane rocket stage
1997	Inactive NOAA 7 satellite hit by un-cataloged debris large enough to change its orbit and create additional debris
2002	Inactive Cosmos 539 satellite hit by un-cataloged debris large enough to change its orbit and create additional debris
2005	US Rocket body hit by cataloged debris from Chinese rocket stage
2007	Active Meteosat 8 satellite hit by un-cataloged object large enough to change its orbit
2007	Inactive NASA UARS satellite believed hit by un-cataloged debris large enough to create additional debris
2009	Active IRIDIUM satellite hit by inactive Cosmos 2251

[8] Weeden, Billiards in Space, The Space Review, Feb 23, 2009. www.thespacereview.com/article/1314.

2.1.4 Collision Reality (2009 +)

A collision between an active IRIDIUM and a dead Cosmos satellite was the watershed event that brought attention to the space debris issue.   Projections show that the cascade of debris population is becoming a real problem. The community now recognizes that space debris reduction must be pro-active, not simply passive. The major concern is that without restrictions in the growth of space debris (or reduction of total numbers), collisions amongst the debris will increase dramatically leading to a potentially unmanageable growth of in-orbit items.   This environmentally catastrophic effect, called the Kessler Cascade, is shown in Figure 2.1. This cannot be allowed to occur; and, efforts must be initiated to reduce debris growth through mitigation and active reduction.

Initiation of a space elevator and its survivability will be enhanced when large inert objects are actively removed.   Many knowledgeable professionals believe that space faring nations must remove at least five to ten large rocket bodies and spacecraft from orbit each year. The growth of space debris will not be slowed or stopped by current space faring nations' passive mitigation techniques. Indeed, Figure 2.1 shows potential run-away growth even with the impossible assumption of zero future satellite launches -- Ever. A recent recommendation from NASA is that at least five large objects be removed from LEO each year to slow down the growth of space debris. We believe there must be a more aggressive approach, perhaps 25 per year, to improve the situation and significantly lower the danger of small debris hitting large objects and causing explosions resulting in ever increasing numbers of dispersed space debris. This pamphlet recognizes that NASA would like to remove about 2,000 large spacecraft and rocket bodies from the space debris catalog. This would significantly lower the probability of a future Kessler cascade syndrome; however, the authors of this study will make the assumption that only modest successes will occur and debris will continue to grow - especially in LEO.

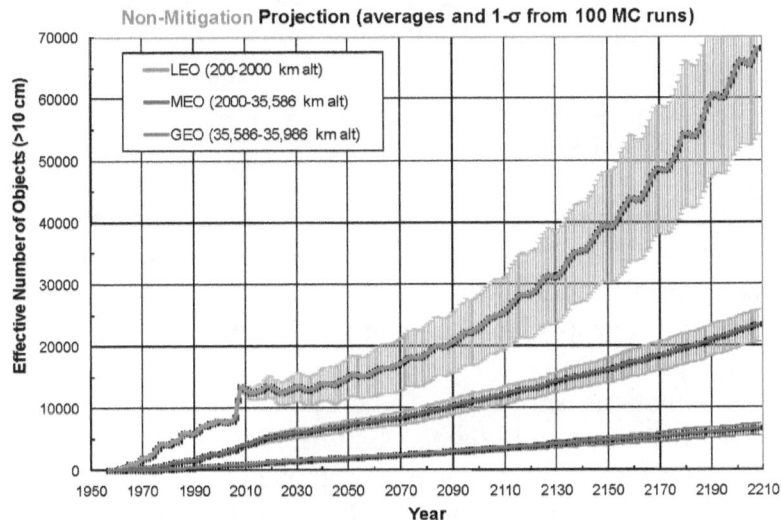

Figure 2.1 Potential Growth Patterns – Kessler Cascade[9]

2.2   New Century's Environment

The first decade of the 21st century ushered in a new environment that is directly applicable to both the space debris and space elevator communities. The scientific and engineering communities changed the way they thought about space. The watershed event was the collision between an inactive Cosmos 2251 satellite with the active IRIDIUM #33; and, it focused people's attention on the space debris problem.

Two other events involved the deliberate destruction of satellites. The first was an anti-satellite demonstration by the Chinese exploding the Feng Yun 1C, which generated an additional 3,000 pieces of debris (mostly above the International Space Station).

---

[9] With permission from Debra Shoots, NASA Orbital Debris Program Office, May 2010.

The second event was when the United States shot down one of its own dead satellites. The exercise was designed to leave no debris in orbit from either the missile payload (on sub-orbital path) or residual debris from the target spacecraft. Almost all debris (well over 95%) decayed within a year, while being significantly below human spaceflight altitudes. With knowledge of debris particles currently in orbit, the calculations were run to assess the situation. The answer came back in two parts:

Answer 1: We have abandoned the philosophy of doing nothing – the Big Sky theory is no longer applicable as a policy. Space faring nations MUST act in more than a passive manner if LEO is to be available to us in the future.

Answer 2: Calculations show that the orbital environment is fragile and actions must be initiated. Estimates indicate that a minimum of five large objects must be removed each year to slow the growth problem we have today. Although a big rocket body only counts as a single piece of debris, it has the potential to fragment into thousands of pieces of debris when hit. There are over 2,000 large pieces that need to be removed to ensure the cascade effect does not define the future environment for near Earth orbits.

These two answers were discussed at the December 2009 conference, sponsored by NASA and DARPA in Washington D.C. Papers described the problem and explained the physics of collisions; however, very few actually addressed "how-to" remove debris from orbit. The papers and discussions showed that there must be an immediate implementation of a space debris removal plan as well as improved tracking and conjunction analyses.

2.3 Space Debris Description

What is the probability of puncture from impacts of small debris? What is the probability of severing by large orbiting objects? How should the space elevator community plan to mitigate these threats? This pamphlet divides the problem into altitude regions to demonstrate why the LEO environment is where the greatest hazards exist; that the MEO region has a low threat environment [along with Super GEO]; why GEO has slowly drifting space debris; and, how the atmospheric region does not present a debris issue as space systems do not spend significant operational time below 200 km. This pamphlet addresses three debris threat categories: (a) small (less than 10 cm) which are numerous (10 times the tracked numbers) with random direction, (b) tracked and inert (10 cm and larger) with known numbers and orbital characteristics, and (c) large and controllable (active satellites are about 6% of tracked). During this discussion, the basic assumptions are:

- Knowledge of the space elevator incremental segment locations will be estimated from known measurements (GPS, radar, ribbon riders, predictions, and retro reflectors).
- Knowledge of the debris environment will be at least to today's knowledge base [cm's for exceptional satellites, meters for many large satellites with GPS, 100's of meters for most, and 10's of kilometers for some].
- Only six percent of tracked orbital items are under control with predictable movements, enabling them to maneuver around a space elevator.
- Current and future space faring nations will improve their debris mitigation programs over the next ten years.
- Some type of active removal will be initiated in the next ten years to ensure the Kessler Cascade does not occur.

Note on Debris Management: The current debris regime views the locating of each item in space as a government function. The future demands a move from embryonic national level systems to global space tracking and management systems. Space traffic management should become an active international responsibility.

2.4 Space Debris Population

After over 50 years of space operations by numerous international players, more than 35,000 objects have been catalogued with over a third still in orbit. The table to the right depicts the current (April 2010) population of objects as small as 10-20 centimeters for LEO (200-2000 km altitude) and objects as small as one meter in Geosynchronous Equatorial Orbit

Satellite Box Score			
(as of April 7 2010, as catalogued by the U.S. Space Surveillance Network			
Country/ Organizati on	Payloads	Rocket Bodies & Debris	Total
China	85	3207	3292
CIS	1400	4370	5770
ESA	38	44	82
France	48	421	469
India	39	131	170
Japan	112	77	189
US	1127	3694	4821
Other	463	114	577
Total	**3312**	**12058**	**15370**

(GEO). The minimum object size reflects the capabilities of the US Strategic Command's Space Surveillance Network (SSN). However, the LEO debris problem must be kept in perspective. The density is still quite small as there is only one large spacecraft item in low Earth orbit in each 750 x 750 x 750 km cube and only one small piece of debris (10 cm or larger) in each 90 x 90 x 90 km cube.

The altitude distribution of debris must be understood when dealing with threats to a space elevator. The total length is not really at danger because most altitudes do not have any significant distribution of debris. There is concern at GEO (where the large objects are not going very fast with respect to a space elevator) while at LEO it is much more important to understand the numbers because of increased densities and high velocities. This requires a methodology that addresses the differences in altitude and density.

Figure 2.2 shows the growth in numbers of objects vs. time. The orbital inclination is not really relevant as all items will cross the equator in each of its orbits, no matter what its inclination. The figure has two significant steps in quantity of space debris reflecting the Chinese ASAT test and the IRIDIUM-Cosmos collision.

Multiple sources (optical observations, data gathered in orbit [e.g., Long Duration Exposure Facility-LDEF], and statistical methods) estimate that there may be as many as 100,000 additional objects in earth orbit that are too small to be tracked by the Space Surveillance Network (SSN). An easy estimate relates the known debris distribution with the "small stuff" and multiplies it by ten. [15,370 x 10 = 153,700 assumed at < 10 cm].

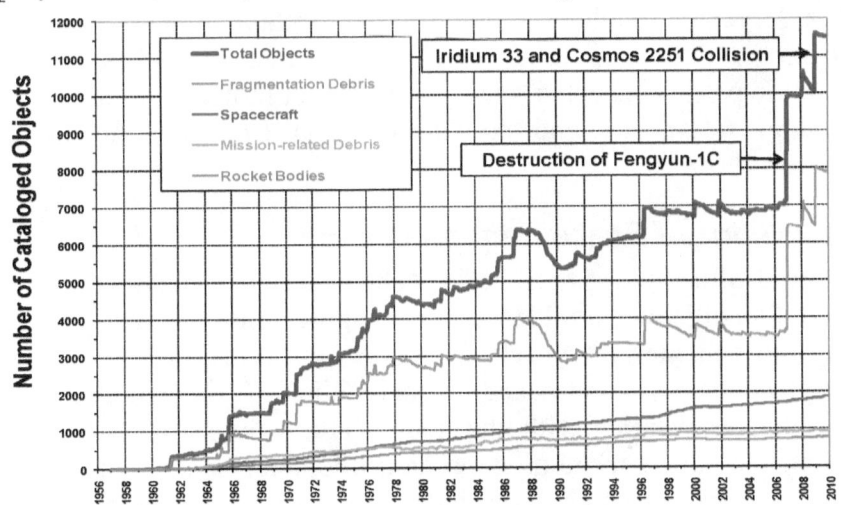

Figure 2.2 Growth in numbers of objects vs. time[10]

This chart shows the dramatic increase due to the Chinese ASAT test and IRIDIUM-Cosmos collision. These numbers show how attitudes about space debris shifted through the four phases (as

10 With permission from Debra Shoots, NASA Orbital Debris Program Office, May 2010

described in section 2.1) and reflect the current and future threat to a space elevator. A good rule of thumb is that the LEO numbers account for slightly greater than two thirds of the total objects. This is the area where we should focus on debris mitigation, such as "taking the hit" or collision avoidance actions. Again quoting from the position paper[11], "Only about 6% of the cataloged objects are operational satellites. About one-sixth of the objects are derelict rocket bodies discarded after use, while over one-fifth are non-operational payloads. Pieces of hardware released during payload deployment and operations are considered operational debris and constitute about 12% of the cataloged population. Lastly, the remnants of over 150 satellites and rocket stages that have been fragmented in orbit account for over 40% of the population by number. These proportions have varied only slightly over the last 25 years. Small and medium-sized orbital debris (size ranging from 1/1000 mm to 20 cm) includes paint flakes, aluminum oxide particles ejected during solid rocket motor booster firing, breakup fragments, and coolant droplets from leaking nuclear reactors."

2.5 Knowledge of Space Debris Location

As noted by Loftus and Stansbery[12] "There are two distinct phases..." of collision avoidance: cataloging objects and maintaining full ephemeris for each. As one would imagine, the accuracy of the ephemeris on tracked objects in the SSN (shown in Figure 2.3) database varies with the source and volume of the observations. Accuracy can be as good as a kilometer or two for objects that are tracked frequently by radar. [Note: the frequency of measurement depends upon multiple factors; size, priority because of threat potential, location of radars, and operational needs.] Less frequently tracked objects can vary from a few kilometers to tens of kilometers. The large majority of catalogued objects have accuracies in the several kilometers to tens of

---

[11] 2001 Position Paper On Orbital Debris, International Academy of Astronautics, 24.11.2000.
[12] Loftus, J.P. and Stansbery, E.G. Protection of Space Assets by Collision Avoidance

kilometers range. The Space Control Center (SCC) is the terminus for an abundant and steady flow of information from the space surveillance network. It has large and powerful computers to store "observations" which include both time tagged optical and radar measurements which sometimes include size estimates in the form

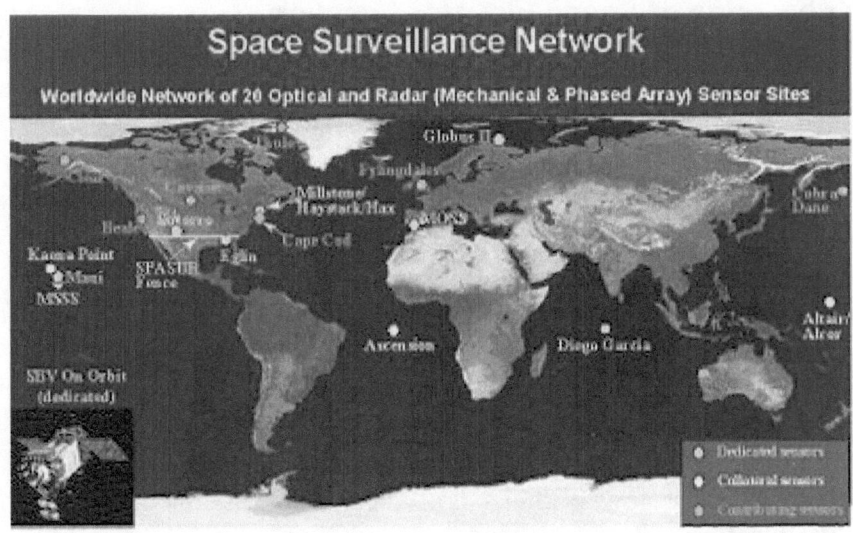

Figure 2.3 Space Surveillance Network[13]

of average radar cross section. The SCC computes and stores ephemeris for tracked objects. Owners of operational satellites may know the locations of their satellites to much better accuracies. For example, Gravity Probe B and Global Positioning System (GPS) satellites are known to within a few, or a few tens, of centimeters with post processing of measurements. IRIDIUM constellation satellites are known to a few tens of meters each. Predicting where catalogued objects will be is a function of the accuracy of the ephemeris (and size estimate) and the accuracy of the propagator [or computational estimator of future location]. The accuracy of the projected ephemeris can be kilometers to tens of kilometers. Propagators, particularly those for LEO, perform poorly because there is still great uncertainty about atmospheric drag,

[13] www.au.af.mil/au/awc/awcgate/usspc-fs/space.htm

Earth oblateness, sun and moon effects, and other factors. The SCC also runs the Computation of Miss Between Orbits (COMBO) software to predict collisions for selected objects such as the US Space Shuttle, which has a keep out zone of 25 km. The US Space Shuttle has used these predictions to maneuver out of harm's way several times.

2.6   Knowledge of Space Elevator Location

By employing GPS receivers at multiple locations on the ribbon, taking measurements frequently, and utilizing powerful computers (Kalman filters), we would expect our knowledge of the location of the ribbon to be in the meters to tens of meters. However, because the location of the elevator is critical to any mitigation technique, and simplicity is an essential trait of ribbon design, another natural solution presents itself. Small or flexible (so they do not interfere with the cargo carriers) corner cube reflectors could be placed along the ribbon during deployment. An automated triangulation system could be established to estimate ribbon location to within centimeter accuracies. Three lasers (with backup, of course) would be sufficient to irradiate each designated ribbon segment every few minutes and register exact distances from known locations. With knowledge of position over time, the motion of the ribbon itself will be more accurate and easier to predict.

2.7   Position Problem

As discussed in this chapter, the knowledge of an individual body includes its location when measured and its propagation into the future. Armed with knowledge of a close approach and probability of collision, owners can take actions to maneuver if they are able (and if they choose to do so): however, this may degrade mission performance. This leads to two parts of the space elevator survivability space debris problem;

- Probability of collision: This is based upon probability theory, density of debris, and cross sectional areas of target (space elevator).

- Maneuvering of the space elevator: A subset of the issue deals with the uncertainty of knowledge and the location of the space elevator. This determines "how far the space elevator has to be moved?"

One conclusion is that better knowledge of location leads to more confidence in projected conjunctions between space debris and a space elevator.

Chapter 3 – Probability of Impact

3.0   Determining the Probability

The probability of collision between a space elevator and space debris requires the consideration of many variables. They range from the actual population, or density, of the space debris to the velocity difference between the debris and the space elevator, to the amount of time between collisions. NASA's Orbital Debris Program Office provided the data in altitude chunks of 20 km lengths for this analysis. Each of these represents, in the probability of collision calculation, a spherical shell with the appropriate number of debris across a 20 km length of the space elevator.

3.1   Density of Space Debris (by altitude region)

It is important to estimate the densities of known and estimated [unknown] space debris to calculate the collision risk. Figure 3.1 shows densities of space debris per unit volume by altitude and is used to calculate the probability of collision.

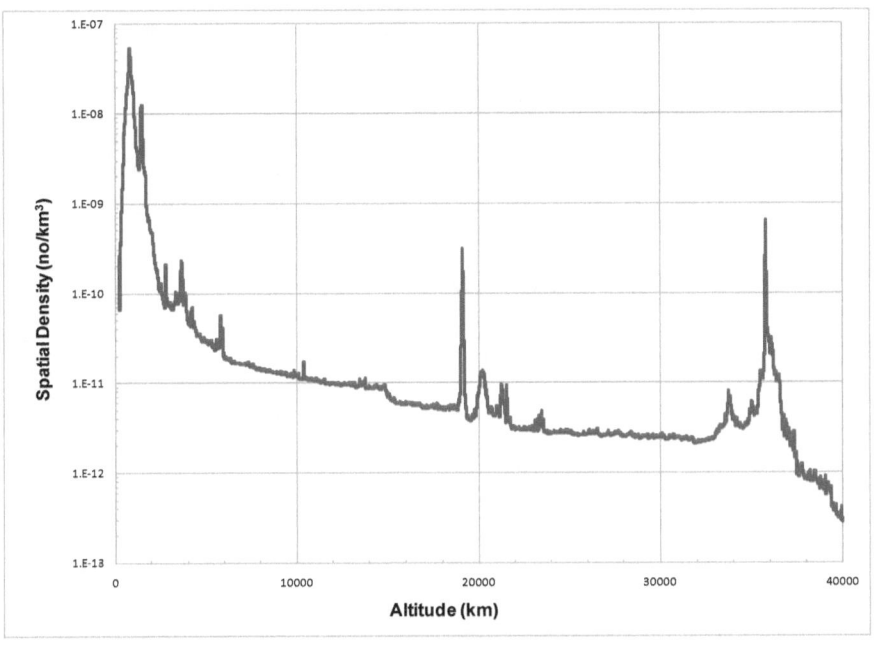

Figure 3.1 Spatial Density[14]

3.2   *Relational Velocities*

Determining the probability of space collisions actually requires three different calculations: head-to-head approaches, tail catch-up collisions, and oblique, or orthogonal, impacts.   Each has a different set of approach velocities and areas of potential collisions.   This perspective of debris to debris collision is important when considering a space elevator.   The big difference is that space elevator velocity varies linearly by altitude, not by orbital equations.   At the surface of the Earth, on the equator, the linear velocity of the anchor is 0.48 km/sec [or 360 degrees rotation in one day with the circumference of the Earth].   Each elevator segment has a unique linear velocity depending on its radius from the center of the Earth and its constant rotation.   Table 3.1 shows the linear increase in velocity of a space elevator from

[14] *With permission from Debra Shoots, NASA Orbital Debris Program Office, May 2010.*

the surface to geosynchronous altitude. The table also shows the circular orbital velocity for the appropriate altitudes. The difference is then the potential collision velocities at any given altitude. Note: the geosynchronous transfer orbit (or highly elliptical orbit) velocities are also shown at their perigees (LEO region) and apogees (GEO region).

Table 3.1: Velocity Descriptions

altitude	Circular velocity	Elliptical Velocity	Space Elevator Velocity	Elevator Impact Velocity
km	km/sec	km/sec	km/sec	km/sec
200	7.78		0.48	7.8 +/- .48
2000	6.90		0.61	6.9 +/- .6
20200	3.87		1.94	3.9 +/- 1.9
35536	3.08		3.06	+/-20's m/sec
35786	3.07		3.07	+/-10's m/sec
36036	3.07		3.09	+/-20's m/sec
600	HEO perigee	9.90	0.51	9.9 +/- .5
35786	HEO Apogee	1.64	3.07	3.1 +/- 1.6

Note: rare velocity differences at GEO could reach 125 m/sec

Table 3.2 Altitude Regions & Relational Velocities

Region	From (kilometers)	Relational Velocity (km/sec)
Super – GEO	36,036 - 100,000	Elevator Velocity + Asteroid Velocity >> 10
GEO	35,680 – 35,880	Single digits to Tens of meters/second
MEO	2,000 - 35,680 NAV – 20,200	HEO perigee = 9.9 HEO apogee = 1.6 Navigation orbit= 3.9
LEO	Spaceflight limit 200 - 2,000	Normally 6.9 to 7.4 also HEO perigee
Aero Drag	Sea Level - Spaceflight limit	Rapidly decelerating, but still significant

[GEO – geosynchronous orbit @ 35,786 km; MEO – Medium Orbit;
LEO – low Earth orbit; Radius of Earth 6378 km]

3.3   Risk of Debris to Space Elevator

Quoting from the 2001 IAA Position Paper On Orbital Debris[15], "The probability that two items will collide (PC) in orbit is a function of the spatial density (SPD) of orbiting objects in region, the average relative velocity (VR) between the objects in that region, the collision cross section (XC) of the scenario being considered, and the time (T) the object at risk is in the given region.

$$PC = 1 - e^{(-VR \times SPD \times XC \times T)}$$

This relationship is derived from the kinetic energy theory of gases which assumes that the relative motion of objects in the region being considered is random." This methodology was introduced in 1983, by Penny/Jones in their Master's thesis "A Model for Evaluation of Satellite Population Management Alternatives.[16]

---

[15] 2001 Position Paper On Orbital Debris, International Academy of Astronautics, 24.11.2000.

Note, that the PC equation may be approximated by the product of the four terms as long as the value is very small (less than 1/100). As the cataloged population, lifetime, and satellite size increase, the PC will also increase. We do not use the product method if we anticipate the PC being larger than 1/100. An example of area is (if we consider the LEO area [200 to 2,000 km altitude] of the ribbon) the cross sectional area 1,800,000 meters times 1 meter or 1,800,000 square meters, or 1.8 square kilometers. The relative velocity is the average velocity for the orbiting objects. In LEO, there are tens of thousands of tracked objects, so the calculation leads to valid estimates.

3.4   Probability of Collision (PC)

The probability of collision can be broken into separate illustrative cases. This pamphlet sets up the representation of several cases by altitude region [LEO cases A, B, & C; MEO case D; GEO case E] as identified in altitude density shells. In the LEO orbital region, two shells are 60 km in thickness and represent the area where the tracked space debris is most dense [Case A] and average [Case B]. In addition, a third case in LEO deals with all the debris from 200 to 2000 km altitude [Case C]. Another dimension for the description of LEO cases is the "untracked" (estimated) density [Cases A-u, B-u, C-u] where the numbers are estimated to be ten times the tracked numbers inside each case. A third dimension is the representation of operational spacecraft which can maneuver as they are still being controlled by the ground [Cases A-c, B-c, and C-c]. Operational spacecraft numbers are assumed to be six percent (0.06) of the tracked space debris. Case D represents MEO while Case E represents GEO. The cases are shown below:

---

[16] Penny Robert and Jones, Richard, "A Model for Evaluation of Satellite Population Management Alternatives," AFIT Master's Thesis, 1983.

Low Earth Orbit (9 cases)

Case A: 60 km ribbon segment (740-800 km altitude) representing the peak debris density – highest risk case.

Case B: 60 km ribbon segment (1340-1400 km altitude) representing an average debris density in LEO.

Case C: 1800 km ribbon segment (200-2000 km altitude) representing the entire LEO environment.

Case A-u, B-u, C-u: represent the untracked items in above described segments. Estimated to be ten times the tracked debris.

Case A-c, B-c, C-c: represent the controlled satellites in above segments. Estimated to be six percent of the tracked debris.

Medium Earth Orbit (1 case)

Case D: 200 km ribbon segment (around 20,200 km altitude) representing the navigation orbit environment [only tracked items are calculated].

GEO Orbit (1 case)

Case E: 200 km ribbon segment (35,680 - 35,880 km altitude) representing the GEO environment [only tracked items are calculated].

As we noted earlier, the probability of collision is a function of the relative velocity (VR), the density of objects (SPD), the cross sectional area (XC) and time (T). This approach works well for LEO where the behavior of Earth orbiting objects is very similar to the behavior of gas molecules (as noted in Section 3.1). It is less similar for MEO and GEO; however, we use the same methodology as we lack anything better. We will use the formula PC = 1 – e(-VR x SPD x XC x T) for all eleven cases.

3.4.1 LEO Cases

The three baseline cases for LEO tracked debris will be run for the probability of collision (PC) in LEO for three threat types: untracked (< 10 cm), tracked (> 10 cm), and cooperative satellites. This range of altitude segments and debris types attempts to layout the range of threats that a space elevator will encounter in day to day operations in Low Earth Orbit. This requires a total of nine cases for LEO predictions of collision.

Table 3.3 LEO Regional Breakout by Cases

Types of Debris	Case	Comment
Untracked Debris < 10 cm		10 x tracked
60 km stretch - peak	**A-u**	Highest Density
60 km stretch - average	**B-u**	Average LEO
LEO 200 - 2000 km	**C-u**	Total LEO region
Tracked Debris > 10 cm		
60 km stretch - peak	**A**	Highest Density
60 km stretch - average	**B**	Average LEO
LEO 200 - 2000 km	**C**	Total LEO region
Cooperative Debris		0.06 x tracked
60 km stretch - peak	**A-c**	Highest Density
60 km stretch - average	**B-c**	Average LEO
LEO 200 - 2000 km	**C-c**	Total LEO region

The numbers of objects tracked in LEO are illustrated by Figure 3.2 from NASA's Orbital Debris Program Office. As seen, there is a peak at 740-800 km. In addition, if you do the "eye-ball" smoothing across 200-2000 km, 1340-1400 km reflects an average density. As reflected in this chart, two-thirds of all tracked debris are between 200 and 2000 km in altitude. The following table (Table 3.4) shows the significant case of tracked space debris with the calculated probability of collisions between LEO tracked debris and a space elevator.

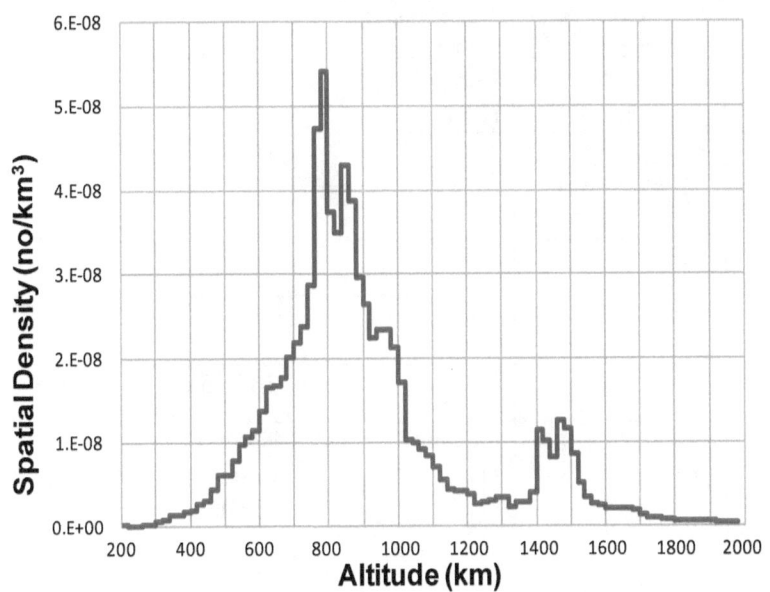

Figure 3.2 LEO Spatial Density[17]

[17] *With permission from Debra Shoots, NASA Orbital Debris Program Office, May 2010.*

Table 3.4 Probability of Collision for Tracked Objects

	PC Case A 60 km all tracked, (peak band)	PC Case B 60 km all tracked, (LEO avg)	PC Case C LEO all tracked objects
Top Altitude	800	1400	2000
Bottom Altitude	740	1340	200
Tracked Objects	1672	149	11298
Ribbon Area (km²)	.06	.06	1.8
Time (days)	365.25	365.25	365.25 (1)
Probability of Collision	**0.457859 per year**	**0.043647 per year**	**0.969317 per year (.00949 per day)**

- **Case A** Results show that the tracked items have a one-in-two chance of having a conjunction with the space elevator each year across a 60 km segment in the high threat region. Very limited number of space elevator 60 km segments across LEO are at high density risk levels.
- **Case B** Shows that the average in LEO, for any 60 km segment, is around one-in-twenty chances per year [most LEO 60 km segments have less].
- **Case C** The full spread across LEO shows the probability of conjunction for tracked objects is essentially three per year. This means that some location across 1,800 km will have a potential conjunction by tracked debris every four months.

One must remember that, when dealing with tracked objects, we know where the debris is and can predict its future location to enough precision to enable us to make judgments as to the specific risk per opportunity for conjunction.

Similar calculations were conducted across three cases for the "small stuff," or untracked debris. The summary probability of collision for a space elevator with untracked objects is shown in Table 3.5.

Table 3.5 Probability of Collision for Untracked Objects

	PC Case A-u 60 km all un-tracked, (peak band)	PC Case B-u 60 km all un-tracked, (LEO avg)	PC Case C-u LEO all un-tracked objects
Top Altitude	800	1400	2000
Bottom Altitude	740	1340	200
Tracked Objects	16720	1490	112980
Ribbon Area (km^2)	.06	.06	1.8
Time (days)	365.25 (1)	365.25 (1)	365.25 (1)
Probability of Collision	**0.9978 per year (.0166 per day)**	**0.3500 per year (.00122 per day)**	**0.9999999 per year (.0949 per day)**

This second type of debris is the untracked set, which was earlier estimated to be roughly ten times the density of the tracked set. With this as the starting position, the probability of conjunction (PC) for:

        Case A-u is one-in-60 days,
        Case B-u is one-in-700 days, and
        Case C-u is one-in-ten days.

The cross-sectional area of the untracked space debris is less than 10 cm (with the preponderance much smaller) which, because of its velocity difference, should just "blow through" the ribbon when it actually collides.

To put this in perspective, if you were to look at the probability of collision for one square meter of space elevator ribbon (1 m wide by 1 m long) in Low Earth Orbit (200-2000 km altitude), the probability is about once every 2,000 years for any specific ribbon square meter. As the danger area is the full LEO environment (200-2000 km length), the summation of these probabilities for each of the 1,800,000 meter squares is equivalent to once every ten days. However, the probabilities of multiple impacts on any single square meter of ribbon are extremely small!

The third type of debris is the tracked and cooperative set, which includes all operational satellites in Low Earth Orbit. Space elevator operators will track operational satellites; and, then, work with the owner as to appropriate actions to ensure collision avoidance. This is beneficial to both parties. As this is approximately 6 % of the tracked debris, the probabilities are as follows:

> **Case A-c** yields a collision every 30 years
> **Case B-c** yields a collision every 400 years
> **Case C-c** yields a collision every 5 years

The summary of the probability of collision for space debris with a space elevator in the LEO region is summarized below:

Table 3.6 Probability of Collision LEO Summary

Types of Debris	Case	Probability of Collision
Untracked Debris < 10 cm		PC per day
60 km stretch - peak	A-u	1.66%
60 km stretch - average	B-u	0.12%
LEO 200 - 2000 km	C-u	9.54%
Tracked Debris > 10 cm		PC per year
60 km stretch - peak	A	45.79%
60 km stretch - average	B	4.36%
LEO 200 - 2000 km	C	96.93%
Cooperative Debris		PC per year
60 km stretch - peak	A-c	Every 30 yrs
60 km stretch - average	B-c	Every 400 yrs
LEO 200 - 2000 km	C-c	Every 5 yrs

3.4.2 Medium Earth Orbit Case

The medium altitude orbit covers a range of mostly empty space. The region around a 12 hour circular orbit for navigation satellites is populated (estimate 200+ satellites) with circular orbits of eight to twenty satellites per orbital plane; however, the good news is the volume is huge. The spherical shell (200 km in radius height) centered at 20,200 km altitude is labeled as case D. Another case is the geosynchronous transfer orbit with rocket bodies and satellite residuals. This orbit has a lot of residual rocket bodies; but, they are numerically not a large threat because of the vast volume and the location of perigee. This case is described as:

Case D: 200 km ribbon segment (20,200 km altitude) representing the Navigation orbit environment. (see Table 3.7 for results)

3.4.3 GEO Case

The GEO belt is extremely interesting and has many operational spacecraft generating large profits for commercial enterprises. As such, the space elevator must not interfere with GEO operational satellites. In addition, derelict spacecraft in this orbit are all going in the same direction as the space elevator. This means that the likelihood of fragmentation of these satellites or damage to the tether is greatly reduced. This case is described as:

Case E: 200 km ribbon segment (35,680 - 35,880 km altitude) representing the GEO environment. (see Table 3.7 for results)

Table 3.7 Probability of Collision for Tracked Objects

	PC Case D 200 km all tracked, at MEO	**PC Case E** 200 km all tracked, at GEO
Top Altitude	20300	35880
Bottom Altitude	20100	35680
Tracked Objects	22	600
Ribbon Area (km^2)	0.2	0.2
Time (days)	365.25	365.25
Probability of Collision	**0.00030 per year**	**0.0026 per year**

The probabilities of collision are as follows:

Case D is 3 in 10,000 years
Case E is 3 in 1,000 years

3.5   Summary Probability of Collision

After evaluating all of the eleven cases, the numbers show that LEO is the highest threat arena. We know this intuitively as the density of space debris is greatest at LEO and it has the highest differential velocities – two major drivers in the probability of collision equation. In addition, as the population density is not great at MEO and the volume is huge, MEO still falls into the "Big Sky Theory" of less worrisome. The GEO orbit has a restrictive band [Sir Arthur Clarke's altitude for station keeping at zero latitude] of limited population. This leads to some concern from the numbers; however, the differences in velocities are so small that the danger is even smaller. The next chart summarizes the concerns for all eleven cases.

Table 3.8 Summary of Probability of Collisions

Types of Debris	Case	Collision About Every
Untracked <10 cm		
60 km stretch peak	A-u	60 days
60 km stretch average	B-u	2.5 years
LEO 200-2000 km	C-u	10 days
Tracked Debris >10cm		
60 km stretch peak	A	2 years
60 km stretch average	B	23 years
LEO 200-2000 km	C	1.3 years

Cooperative Objects
60 km stretch peak	A-c	30 years
60 km stretch average	B-c	400 years
LEO 200-2000 km	C-c	5 years

Tracked Debris >10cm
200 km stretch-MEO	D	> 4000 years
200 km stretch-GEO	E	400 years

These results lead us to the following conclusions:

GEO altitude belt is not a problem.

MEO volume is not a problem.

Untracked, small (<10 cm debris) will, on the average, impact a Space Elevator in LEO (200-2000 km) once every ten days, and therefore, the ribbon must be designed for impact velocities and energies.

Tracked debris will impact the total LEO segment (200 – 2000 km) once per 100 days or multiple times a year if no action is taken.

Tracked debris will, on average, impact a single 60 km stretch of LEO space elevator every 18 years and every five years in the peak regions if no action is taken.

Chapter 4 – Mitigation Techniques

4.0   User Needs – System Objectives

A space elevator must be designed with many factors included in the trade space.   Some anticipated issues for proprietors, customers, and the space community as a whole, are centered around survivability of the architecture vs. space debris:

- Zero Sever Infrastructure – the space elevator, once established, must never be severed. The Human Race must never be dominated by the gravity well again.
- Robust Ribbon – the ribbon must be able to withstand small debris "hits" and keep on operating.
- Robust Situational Awareness –   knowledge of the environment must be as complete as possible – better tracking of space objects and location of space elevator segments, as well as much better propagation techniques for predicted ephemeris.
- Multiple ribbons –   ensuring the continuation of "winning the gravity well wars."

These objectives lead to a basic expectation of space elevator infrastructure:

"Safe Operations"

4.1   User Requirements

The following user requirements cover many issues within the Space Elevator Safe Operations Concept. A quick summary is shown in Table 4.1 with many of the items directly related to the problem of space debris.

Table 4.1 Performance Requirements

Basic	Detailed Requirements
Zero Sever	Multiple ribbons. Single catastrophic severing of a space elevator ribbon must not disable rocket free access to space.
	Low occurrences of lightning damage
	No explosions on ribbons
	Low occurrences of high winds/hurricane damage
	Laser power support does not melt ribbon
	No orbit/fly/float/drive within the space elevator corridor
	Debris/meteoroids tracked and predicted
	Robust ability to move ribbon to avoid major space debris
	Ability to move ribbon to avoid spacecraft, as needed
Robust Ribbon	Safety factor of 2.5
	Tolerance for atomic oxygen
	One-meter wide ribbon, curved for multiple hit (small particles) tolerance
	Tolerance for bending modes
	Tolerant to climber forces
Robust Situational Awareness	Knowledge of solar/lunar effects (ultra violet, 7 hour oscillation, radiation)
	Tracking of satellite/rocket bodies
	Tracking of space debris
	Leadership in global debris mitigation efforts
	International policy creator/enforcer
	Enabler of debris reduction
	Knowledge of space elevator segment location

4.2   Potential Solutions to Debris Threats

These requirements necessitate an understanding of the varying characteristics of space debris and its impact upon space elevators. Such an understanding leads to potential solutions in mitigating the threat. The following approaches synthesize these characteristics and offer engineering solutions that will allow for the safe operation of a space elevator. The following are described in general terms and may be applied along the total ribbon length or where a localized threat is most significant/prevalent.

4.2.1   Do Nothing

Space elevator operators most likely will decide to do as virtually all satellite operators do -- accept the risk. It is possible that a ribbon could survive multiple collisions and tolerate the damage. The probabilities of collision are in a space elevator's favor for large objects, acceptable for small tracked items, and expected for the multitude of miniscule objects that will blow through the ribbon. In parallel with the development of the space elevator, the space faring nations should instigate programs to remove tracked debris objects from space. There are attractive concepts[18] that could remove over 2,000 medium to large objects from LEO within a seven year time frame, resulting in a much friendlier environment for all space operators.

4.2.2   Repair Robots

As the small stuff will be hitting a space elevator every 15 days along an 1,800 km ribbon segment in LEO, repair robots should be developed to scout, notice holes, stop and repair holes, and continue inspections. This could be accomplished on a schedule of once a year or so, depending upon actual damage experienced.

---

[18]   Pearson, Jerome, Eugene Levin, John Oldson, and Joe Carroll, "EDDE: ElectroDynamic Debris Eliminator for Safe Space Operations," 13th Annual FAA/AIAA Commercial Space Transportation Conference, Arlington, VA, 10-11 February 2010.

4.2.3 Move Space Elevator

Even though the distances needed to avoid a potential collision might seem large, they are well within our reach. Reeling out just a few meters of ribbon from the terminus host can impart tens of kilometers of lateral distance. Looking at an altitude of 600 km:

- 10 meters spooled out at the end mass results in a little over three km lateral movement
- 100 meters spooled out results in about 11 km of change
- 1,000 meters spooled out results in about 35 km of movement

The method of controlling the direction of movement (normal to the velocity vector of the collider) will be determined during the design process. The most probable approach will be to let the ribbon lag more or less in the direction of Earth's rotation. In addition, the anchor will be mobile and could initiate resonance "waves" along the ribbon with predicted movement of appropriate segments at precise times.

Figure 4.1, Ribbon Concept

4.2.4 Maneuver Collider(s)

As the location of the ribbon will be well known, satellite owner / operators (whose satellites are maneuverable) could maneuver their satellites to avoid a collision.

4.2.5 Ribbon Design

The ribbon design refers to the analyses of various ribbon descriptions with respect to their ability to survive multiple hits over the ribbon's lifetime (from even the smallest meteoroids and space debris). The most obvious threat is from large numbers of small items (less than one cm in diameter); thus - - the survival of a space elevator must allow multiple hits per segment of ribbon over its lifetime. The principle sources of these particles are meteoroids and debris fragments. One possible approach to mitigate this threat is to manufacture a ribbon that is tolerant to punctures. A current design is given in Figure 4.1. Another concern is the phenomenon of hyper-velocity impact with ribbon strands. How is the energy transferred? Does the large energy impact spread out across the ribbon, or is it localized? Can we design the ribbon to gracefully degrade at those impact velocities? This issue can be better understood and solutions evaluated through testing.

4.2.6 Debris Reduction – Policy

The outdated belief that we can continue to operate with minimum debris reduction policies must be changed to a more responsible use of our space environment. The first steps were taken in 1998 with the approval of the Inter-Agency for Space Debris Coordinating Committee (IADC) and the International Academy of Astronautics (IAA) published[19] approach for debris mitigation. Major space faring nations are indeed incorporating space debris mitigation techniques; but, in a modest way. It is good for the world space community in the long run; but, it must be mandated to be effective. Many steps have been implemented over the last ten years by a small group of space debris mitigation experts resulting in a safer environment because of their pioneering efforts. This must be continued, and re-enforced, to ensure that no more rocket bodies fragment; no more satellites are left in their operational orbits after their mission lifetime; and, that no satellites create multiple smaller pieces. Current thinking inside the international debris community is that a policy could be implemented, and enforced, for "Zero Debris Creation." Below are policy candidates for safer near Earth orbits.

---

[19] *Position Paper on Orbital Debris. International Academy of Astronautics, Paris 2000.*

Shared Accurate Ephemerides

Safer near Earth orbits for a space elevator require accurate ephemerides of all space objects to support collision avoidance calculations. The accuracy of an ephemeris should be the subject of further analyses; however, space vehicles with GPS receivers will be very likely to meet any reasonable requirement. The ephemeris should be made available to space elevator operations and to any "clearing house" that might evolve. Both would be dedicated to collision avoidance calculations and providing warning of conjunctions to all owner/operators.

Mandatory De-Orbit Plans that are REAL

Safer low Earth orbits, with a space elevator, require acceptable de-orbit plans. Acceptance criteria for these plans should be the subject of further study; but, existing NASA documents are a good starting place. The IRIDIUM procedures and practices would also be a good source. The plans could include use of the space elevator "tug" for raising and lowering orbital bodies. The plans would also include actions that the space vehicle would take autonomously if loss of command and control capability occurred.

Mandatory De-Orbit Services Purchase

Safer low Earth orbits, with a space elevator, requires satellite operator riders to purchase de-orbit services even when their de-orbit plans have been accepted. This will mitigate against loss of command and control capability. If the owner/operator successfully performs the end-of-life de-orbit, the fees collected for de-orbit services would be returned.

Mandatory "Safe Mode" Features

Safer near Earth orbits, and a space elevator, could also require space vehicles to have "safe modes" that could place the spacecraft in a "capture friendly" state when loss of contact with the ground occurs. This state can include functions to perform vehicle safeing and removal of all stored energy. Propellant can be used to de-orbit or re-orbit to a more capture friendly orbit. The safe mode must also include a state for operations while riding on a space elevator as well as operations near it. Verification of these features would occur by reviewing design documents [to include the major design reviews]. High fidelity simulation would also be used to verify functionality.

Mandatory "Capture Friendly" Features

Safer near Earth orbits, with a space elevator, require that space vehicle riders have "capture friendly" features. The most important feature will be a standard interface to a space elevator tug which should also be the standard interface for all capture vehicles. The interface would be used for normal orbit raising and de-orbit services enabled when the client space vehicle is 3-axis stable. Verification of these features would occur by reviewing design documents including the major design reviews.

Required Space Elevator Friendly Orbits

Safer low Earth orbits, with a space elevator, requires LEO satellites to be inserted into either a repeating ground track [not coincide with space elevator location] or a capability to maneuver out of the space elevator safe corridor [vertical cylinder one km diameter].

4.2.7 Debris Reduction – Elimination

To increase the probability of survival of a space elevator, the number of large rocket bodies and dead satellites can be "controlled." This concept has at least three approaches:

- grab and de-orbit for low Earth orbiting large bodies
- grab and maneuver as needed for higher orbits
- grab and use GEO belt debris as the space elevator counterweight

The issue is similar in all cases; the inert body must be tracked, rendezvoused with, and captured prior to any action. Many designs have been proposed for this operation. A current concept is capture by a net that is "tossed" over the debris. The net would attach itself to the object/debris easily. The next step would be to stop the inert body's rotation in order to gain control for any action. To stop the rotation, angular momentum must be minimized through an interaction with another force. One idea is to have large balloons (with torque rods) at the end of the ropes to add moment arms and drag. Once stabilized to a certain level, a long tether can be deployed to further stabilize and interact with the magnetic field lines of the Earth for de-orbit drag force creation. At LEO, the length of the tether can be relatively short (10s of km) for rapid decay while at MEO and GEO, longer tethers with weaker forces would result in longer times for desired outcomes. LEO bodies could be burned up; MEO bodies could be placed in space elevator compatible orbits for storage; while, GEO objects could be moved into a location where the mass can be changed from dangerous (crossing the space elevator vertical space corridor) to useful by making it part of a space elevator counterweight beyond GEO. For smaller junk in orbit, many alternatives exist. These include:

- Energy exchange lasers that slow the junk down through "blow-off,"
- Sweepers picking up small things going in a common direction, and,
- Bumper cars for exchange of momentum

To accomplish this task of elimination of junk in space, space nations could fund the clean-up as they do environmental spills. If a space elevator is going to cost in the range of $10-40 billion, a billion dollars could be put forth to clean-up space. How many entrepreneurs will surface when you explain that they can make $100 per kilogram for inert spacecraft or rocket body de-orbits, or movement to a stabilized orbit. This would be roughly 11,000 pieces for $1 billion. Two recent papers[20&21] discussed the concept of docking with space objects and moving them.

4.2.8 Satellite Control – Knowledge

Current radar and optical technology (combined with older computers and software) leads to a situation where the lack of knowledge of space debris location is worrisome for both satellite operators and space elevator designers. To apply techniques that could greatly enhance the safety of Low Earth Orbits with a space elevator, precise and on-going knowledge of orbiting particles is needed. New emphasis must be applied to better tracking (maybe even from platforms on a space elevator), computing, understanding, and prediction.

4.2.9 Satellite Control – Maneuver

As a space elevator is developed, new spacecraft should have non-threatening orbits, or, if necessary, maneuver around the vertical space corridor of a space elevator. This would require a more robust propulsion system with controls necessary to avoid the vertical space corridor.

4.2.10 Rules of the Road, Nodal Control

In addition to knowledge of where active spacecraft are, there should be a policy at the international level that mandates repetitive orbits well clear of a space elevator vertical space corridor. These are also called harmonic orbits because the periods of the orbits are divisible by an even number and have repeating equatorial node crossing. Most low orbiting satellites have periods

---

[20] Pearson, Jerome, Eugene Levin, John Oldson, Joseph Carroll, "EDDE: ElectroDynamic Debris Eliminator for Active Debris Remova,", International Converence on Debris Removal, Chantilly VA, 8-10 December 2009
[21] Ishige, Yuuki & Satorni Kawamoto. "Study on Electrodynamic Tether System for Space Debris Removal." (IAF-02-A.7.04.) 53rd International Astronautical Congress, 2002.

near 90 minutes, 120 minutes, or multiples of those numbers. With proper planning and execution, orbits can be arranged to have precise segments of the sidereal day. This means that these orbits would be able to repeat equatorial crossing and avoid the vertical corridor of a space elevator. This is the current policy for GEO slots (International Telecommunications Union (ITU) allocated) and could very easily be mandated for other orbits. Most missions in space have multiple requirements that lead to orbital selection. By making equatorial crossings repetitive to avoid a space elevator, [an additional requirement in the design trade space] most missions would not be significantly affected.

4.2.11 Ribbon Motion

A space elevator can be moved from its natural position to avoid collisions. The risk of collision is real; and, therefore, requires this capability as not all maneuvering can be mandated for debris. This motion could be modeled during the design phase to ensure that the dynamic stresses are included in material selections and architecture.

4.3   Systems Approach for Survival

A systems approach for the evaluation of the survival of a space elevator enables stakeholders to confidently proceed with the research and development phase of the program. Even though the threat for space elevators is complex and multi-dimensional, designs are flexible across the spectrum of engineering and operations. This systems approach has the objective of minimizing the risk to a space elevator from meteors, meteoroids and space debris. As such, the rest of the chapter shows a proposed prioritization of mitigation approaches for each altitude region. Table 4.2 shows various approaches and sets a prioritization for a systems solution against debris, operational spacecraft, and meteors/meteoroids. The order for the solution set is different for each altitude region because of the resultant system trades between regions vs. threats vs. mitigation approaches.

Super GEO

Priority # 1   Ribbon Design – The principle threat is micrometeoroids. As such, a robust ribbon design resolves most threats, ensuring survival of multiple hits per section per year enabling mission operation success.

Priority # 2   Rules of the Road – The future of Super GEO satellites is going to be significantly different with easy and cheap access to that altitude. As such, the movement of old satellites to graveyard orbits will change to one of capturing old satellites (and, perhaps, using their mass as space elevator counterweights).

GEO

Priority # 1   Debris Elimination – The largest threat is collision with a large spacecraft or rocket body. Collection of GEO satellites not under operational control would significantly reduce the probability of collision. In addition, this collection of mass could aid in counter weighting for space elevators.

Priority # 2   Ribbon Design – The meteorite threat is still significant and must be accounted for with ribbon design. A design must be robust enough to survive multiple hits per year.

Priority # 3   Satellite Knowledge – The GEO arc is not very well tracked because of marginal optical resolution to 37,000 km and needs improvements to determine the threats from smaller components of older satellites. Perhaps, an on orbit sensor coupled with a sensor located on a space elevator could enhance our knowledge.

Priority # 4   Rules of the Road – Strengthen the GEO ITU rules to ensure no lost satellites or out of control inert bodies. Table 4.3 shows current orbital practices from 1997-2002, with only partial success at having satellites end up in this graveyard orbit. Only 22 satellites, out of 75, were in the appropriate drift orbits according to the International Agencies Debris Committee (IADC) report.

Priority # 5   Ribbon Motion – Dormant GEO satellites and high velocity GEO transfer orbit rocket bodies are large enough to sever the ribbon, but can be tracked, predicted, and avoided.

Table 4.2 Systems Approach to Space Elevator Survival

Region	Aero Drag	LEO	MEO	GEO	S-GEO
Kilometers	< 200	< 2,000	> 2,000 < 35,386	> 35,680 < 35,880	>35,880
Threats	Planes, winds aloft, hurricanes, tornadoes, humans	Meteoroids Debris Density highest, Many inclination & altitudes	Meteorite Less dense debris	Meteoroids slow interaction satellite debris	Meteoroids
Methodology	Priority				
Ribbon Design	3	1	1	2	1
Ribbon Motion	4	2	3	5	
Debris Elimination		4	4	1	
Satellite Knowledge		3	2	3	
Rules of the Road	2	5	5	4	2
Corridor Protection	1				

Table 4.3: GEO Re-orbiting Practices[22]

	1997	1998	1999	2000	2001	Total
Abandoned in GEO	5	8	6	5	6	30
Drift Orbit (too low perigee)	5	6	2	4	6	23
Appropriate Drift Orbit (IADC data)	7	7	4	2	2	22
Total	17	21	12	11	14	75

MEO

Priority # 1   Ribbon Design – As the MEO region is just above LEO, and also has a large set of human made debris in the 12 hour orbit, the ability to survive space debris from rocket bodies and spacecraft must be considered.

Priority # 2   Satellite Knowledge – As in the total area of space debris, better understanding of threats is important and can lead to better operational approaches to mitigation.

Priority # 3   Ribbon Motion   –   Dormant navigation satellites and high velocity GEO transfer orbit rocket bodies are large enough to sever the ribbon, but can be tracked, predicted, and avoided.

Priority # 4   Debris Elimination – Larger pieces of debris in highly elliptical orbits, such as the GEO transfer orbit, are a threat and can be de-orbited relatively easily by using atmospheric drag at perigee.

---

[22]   *Hussey, John, ed.,* Position Paper on Space Debris Mitigation Guidelines for Spacecraft, Draft *– International Academy of Astronautics, 2003.*

Priority # 5   Rules of the Road – The MEO orbit is very important for today's navigation systems. As such, there will be multiple constellations at the "half way to GEO" location and large satellites must be controlled in harmonic orbits so they do not cross the equator in the space elevator corridor.

LEO

Priority # 1   Ribbon Design – Space engineers must assume that a ribbon will be impacted by small space debris and meteoroids. As such, the design of a ribbon must be flexible enough to accept monthly (or weekly) hits and still be robust enough to function for its estimated lifetime of 50 years. The design of a ribbon can provide this capability through multiple strands, weave patterns, etc., maximizing longevity under these conditions.

Priority # 2   Ribbon Motion – This combines with situational awareness to enable operational success. One key element in the concept is multiple base legs that can move the bottom of a single strand elevator by simply changing the length of each leg. The dynamics of space elevator motion can be predicted and incorporated with satellite location knowledge to assist in moving out of the way of large, non-maneuverable or uncooperative satellites.

Priority # 3   Satellite   Knowledge   –   Operational approaches must be implemented for a set of debris mitigation techniques. By knowing the orbits of large space objects, a space elevator can be moved as required. To accomplish this, the precise orbital characteristics (accurate ephemeris) of space objects must be known, and shared.

Priority # 4   Debris Elimination – This concept is an idea whose time has come. We must not only stop polluting our environment; but, we must ensure a safe one. This could accurately be described as an "environmental cleanup" activity.

Priority # 5   Rules of the Road – The reality is that LEO satellites are a staple of national missions. An extra requirement in systems design should lead to orbits that are repeatable. As such, they could avoid a space elevator nodal location. An international Rules of the Road agreement can ensure that mission essential orbits can still be utilized, while maintaining a safe space elevator corridor.

Aero Drag

Priority # 1   Corridor Protection – Rules of the Road for vehicles (planes, boats, etc) will ensure that the corridor does not suffer from accidental collisions.

Priority #2   Rules of the Road – This is an extension of priority #1, but applied to the international arena as do both maritime law and aeronautical treaties.

Priority #3   Ribbon Design – The ribbon must be designed for this unique transition from vacuum to sea level pressure. This transition through the various levels of atmospheric pressure will be dynamic and stressful on the ribbon. However, the ribbon must be manufactured with the stated objective of "no failures" in whatever environment it exists.

Priority #4   Ribbon Motion – This mitigation technique will be utilized when there is a predictable hazard that can be defeated by moving the ribbon legs across the surface of the Earth.

Chapter 5 – Conclusions

5.0   Debris Density Reduction

During the preparation for this pamphlet, it became apparent that the community of space debris experts is at a watershed year. They have convinced themselves that there is a need for more robust action than merely mandating mitigation techniques on rocket and spacecraft designs.   At the December 2009 Space Debris Removal Conference sponsored by both NASA and DARPA, the majority agreed that space faring nations must do more than currently required (but unenforced), they must actually remove large debris from orbit.   This was confirmed in Moscow (April 2010 conference) and Paris (June 2010 conference) with discussions on what types and sizes of debris must be removed, how many per year, and finally what impact would that have on the probability of collisions.   The space elevator community endorses those efforts, but would like to encourage further actions to "improve the environment" by reducing density numbers.

5.1   Probability of Collisions.

Earlier in this pamphlet, the probability of collision for a 100,000 km space elevator with the debris density of April 2010 was calculated. Those numbers showed:

- The geosynchronous (and super GEO) region was not a significant threat of collision.
- The MEO region has similarly low probability of collision.
- The LEO region is the area of major concern with the following insights:
  - Untracked, small (<10 cm) debris will impact a Space Elevator in (LEO 200-2000 km), on the average, once every ten days; and, therefore, must be designed for impact velocities and energies.

o Tracked debris will impact the total LEO segment (200 – 2000 km) once every 100 days or multiple times a year if no actions are taken.
o Tracked debris will only impact a single 60 km stretch of LEO space elevator, on the average, every 18 years and every five years in the peak regions.

5.2 Significant Questions:

In the first chapter, a few significant questions were asked to help identify the principle issues. They are represented here with the conclusions from the analyses.

Q. Does space debris cause concern for space elevator?
Answer: YES.
Q. How precisely does one need to know the location of the space elevator ribbon segments?
Answer: Estimate one meter (can be accomplished by GPS or ground based laser reflectors).
Q. How precisely does one have to know the location, and propagated location of large space debris?
Answer: Within 100 meters for 24 hours.
Q. What are the projected levels of concern and what needs to be accomplished prior to operations?
Answer: Knowledge of all tracked debris with improved propagation models and routine knowledge of ribbon location.
Q. How do we mitigate the risk of orbiting debris and satellites colliding with the space elevator?
Answer: Knowledge and planning.
Q. What is the probability of puncture from impacts of small items?
Answer: Close to 100%; therefore, it must be assumed in the design phase of the ribbon.
Q. What is the probability of severing by large orbiting objects?
Answer: Almost zero.

5.3   Conclusion

Space debris mitigation is an engineering problem with definable quantities such as density of debris and lengths/widths of targets. With proper knowledge and good operational procedures, the threat of space debris is not a show stopper by any means. However, mitigation approaches must be accepted and implemented robustly to ensure that engineering problems do not become a catastrophic failure event.

Chapter 6 – Recommendations

6.0   Recommendations

Recommendations are divided into areas where they can be successfully implemented and will significantly improve the survivability of the Space Elevator vs. Space Debris. The conclusions lay out identifiable actions for the various communities.

6.1   Active Player Actions

6.1.1 Space Elevator Community

The space elevator community must lead the way in working with, and guiding, the space community. One of the first items would be to determine the best way to geolocate ribbon elements down to 100 meter segments to within one meter accuracy. The next item is to ensure the design of the ribbon is compatible with the environment. The current robust design is to have a one meter wide, woven ribbon that is tolerant to small debris penetrations [with, of course, a methodology for inspecting the ribbon and repairing in a timely manner]. Operational procedures must ensure that the ribbon element, whose location we know, will not be in the same location, at the same time, as a larger piece of debris (which can be tracked and its location projected).

6.1.2 Space Community

The space community must continue to improve its reporting and tracking of the environment,. They need to identify and implement programs to assure more precise tracking of debris in a timely manner. Another must is to improve the ephemeris propagation technologies so that the timeline for accuracy can be lengthened to a workable timeframe for the commercial world. The inclusion of GPS capabilities on all satellites as well as the ability of each to communicate to an operations center for timely updates of the database should be mandatory. One policy item that could significantly assist in the process would be the publication of the ephemerides of the debris/satellite in a timely [daily] manner. Another item would be a designation of a set of "rules of the road" so that all satellites could let all others know where they are and where they will be in the future (similar to commercial airliners). A current practice that should become mandatory is de-orbit of all LEO satellites within 25 years. And, of course, the biggest item is to immediately initiate a robust program to remove large debris from orbit [maybe ten items per year per participating country].

6.1.3 Satellite Launcher and Operator

Indeed, the operators of both launch vehicles and satellites must treat their environment in a manner that would encourage others to use the resource. If we are to have robust transportation to and from low Earth orbit, safety factors drive us to clean up space debris. In addition, as the number of assets in space increases, the probabilities of accidents, such as the IRIDIUM-Cosmos crash, increases. And finally, we encourage launch operators to consider how they can benefit from the use of space elevators to move payloads on a "real" transportation infrastructure. While they are considering the change, they can contemplate how a space elevator can make their tasks easier, cheaper, and safer, such as the removal of space debris.

6.2 Concluding Thoughts

The risk of collision of a tracked object with the space elevator is low; but, the consequences are high. Therefore, it must be addressed. Three quick thoughts should stimulate more discussions.

6.2.1 Multiple Space Elevators FIRST!

The primary mitigation technique is multiple ribbons. Once we overcome the gravity well we must ensure we always have a ribbon available to build another ribbon. The risk of collision with an untracked object is high but the consequences are low. Periodic "inspect and repair as necessary" by a repair robot should preserve the capability of individual ribbons. By immediately building the second, and then a third, the likelihood of losing operational space elevator access to orbit diminishes and humankind will never again be subject to the constraints of a gravity well.

6.2.2 Another Perspective – Steps Forward

When it comes to the international community, the general rule is that new owner/operators must not interfere with systems already in place (grandfathered). From a debris mitigation standpoint, it should be expected that space elevator owner/operators must not interfere with existing systems. Therefore, a space elevator should not pose a threat to current orbiting satellite systems. If we consider IRIDIUM (66 satellites in the 774-784 km band) and use the aforementioned formula, IRIDIUM would have a .055 PC with a single space elevator for a year. As a maneuver of a couple kilometers would almost certainly disable the use of their crosslinks, IRIDIUM would likely rather not perform such maneuvers. This requires the space elevator community to ensure their planned operations include collision avoidance activities that do not require existing systems to perform collision avoidance maneuvers.

6.2.3 Another Perspective – Enablers

What is currently not affordable in the space debris mitigation and removal will become easily achievable with inexpensive access to space through a space elevator infrastructure.

6.3   *Aggressively Endorse Initiatives*

As a space elevator concept comes of age, with a solid systems engineering program, three timely initiatives dealing with the space debris community are required:

6.3.1 Initiate Space Elevator Corridor

"Rules of the Road" must be initiated to enable a space elevator vertical corridor to exist. Control of nodal passing must be implemented around the world with a mature set of rules ensuring that a space elevator can become a reality.

6.3.2 Initiate a De-Orbit Capability through A Prize Approach

Many papers and engineering concepts have surfaced that deal with elimination of current and future orbital debris. However, cost has always limited these activities to studies without follow-on engineering orbital tests. As a space elevator is funded and goes forward, investment in environmental cleanup should be included in all planning and funding requirements. One idea is to create a prize for the first organization to de-orbit a rocket body with a current estimated lifetime of ten years or more. The prize could be called the "Space Debris Enterprise Award." In addition, rewards can incentivize de-orbiting debris that is hazardous to the future of space elevators. New debris must become at least as socially, and perhaps legally, unacceptable as terrestrial pollution. Another approach is a space superfund as proposed in http://www.popsci.com/technology/article/2010-12/new-report-calls-space-superfund-clean-junk-low-earth-orbit.

6.3.3 Go Beyond a "Zero Debris" Position

The International Academy of Astronautics has published a position paper on space debris.[23] In that paper the Academy takes the position that it is the goal of all space faring nations to create zero space debris within the three important regions. The LEO, navigation constellation ring, and GEO belt are identified. To ensure a healthy space elevator, the concept must be broadened to include all orbits. The mandatory implementation of Zero Debris Requirements would be early in a space systems design for programs prior to their Preliminary Design Reviews. However, the positive impact on a space elevator and other future initiatives will be tremendous. This pamphlet's concept would be to ensure that

---

[23] *Hussey, John ed.,* Paper on Space Debris Mitigation Guidelines for Spacecraft, Draft *– International Academy of Astronautics, 2003.*

zero debris creation is implemented with a new goal of "improving the environment – not simply less pollution."

6.4   Final Recommendation

We hope that this study has raised the awareness of the problem to the space elevator stakeholders and all other users of the near Earth space environment.   Further, we hope that this study will spur action to implement policies and directives to mitigate and reduce the risk of collision.

Appendix - ISEC

Mission Statement: "... International Space Elevator Consortium (ISEC) promotes the development, construction and operation of a space elevator as a revolutionary and efficient way to space for all humanity ..."

The organization of the ISEC is based upon four pillars: Technology, Law, Business, and Outreach. Each of the pillars is headed by a pillar lead, which functions much like a university's department head. Their job is to start initiatives (projects), pursue collaborations, guide project leads and prospective project leads in pursuing their individual projects, and generally increase the activity level of their pillar.

The four pillars are:

- Technical: Investigates the technical aspects of the space elevator and its development, from the material development, to the ribbon riders design and the power approach for the system. This pillar leads all efforts to understand, encourage development of necessary technologies, facility designs and "real world" testing of key elements of the system of systems.

- Business: Currently developing a business-case study, justifying the cost of a Space Elevator. With the baseline of the GEO satellite market, the future funding flows must be shown as larger than the cost of the system.

- Legal: International Space Law will dominant the legal side of the project and is being investigated in multiple ways at the present time.

- Public Relations: Development of a Press Kit and coordination with the public are the current thrusts.

Notice: This study, or position paper, was reviewed by an ISEC Board of Directors selected peer review team. Any opinion, findings and conclusions or recommendations expressed in this report are those of the ISEC and do not necessarily reflect the views of the sponsoring or funding organizations. For more information about the International Space Elevator Consortium, visit the home pages at www.isec.info. Please visit the ISEC website: www.isec.org

www.ingramcontent.com/pod-product-compliance
Lightning Source LLC
Chambersburg PA
CBHW021905170526
45157CB00005B/1978